PICTURES OF ENGLISH TENSES LEVEL 2

Contents

Teaching Notes

Pictures (Side A) / Exercises (Side B)

1	Overview of tenses
2	Present Perfect (with 'just')
3	For / Since / Ago
4	Past / Present Perfect
5	Past Simple / Continuous
6	Instructions
7	Much / Many / A lot of
8	Comparatives and Superlatives
9	1st Conditional
10	Present & Past Passive
11	Describing things
12	Revision

You will find 'Pictures of English Tenses Practice Cards'
ISBN 1 898295 65 4
a colourful, "right brain" resource for further practice

© English Experience, 25 Julian Road, Folkestone, Kent CT19 5HW

TEACHING NOTES

Using **Pictures of English Tenses Level 2**

- Photocopy Side A of the page to provide **oral practice** in the target structure, and Side B to provide **written follow up.**

- The Teaching Notes on the following pages give you **prompt questions** to generate the correct structure for each tense, and also **example conversations** and **answers.**

- **Pictures of English Tenses Level 2** extends the idea of linking tenses with a colour. These were introduced in Level 1 as:
  Present Simple (Dark Blue)   **Present Continuous** (Light Blue)
  Past Simple (Brown)   **Future** (Yellow)

- In Level 2 we add
  Present Perfect (Green)   **Past Continuous** (Light Brown)
  Passives (Orange)

- Use these colours consistently so students quickly become familiar with them.

- The names of the characters in the picture drills, reinforce this association.

- Students now have a "right hemisphere", colour based, visual system for identifying tenses. It is easy to make a 'colour choice' between, say, Present Perfect and Past. *Is it Green or Brown?*
  The teacher can correct by simply holding up the appropriate coloured pen – and the colour prompts the student into self-correction.

The BRAIN *friendly** colour code

1 **Present Simple** – Dark Blue
2 **Present Continuous** – Light Blue
3 **Past Simple** – Brown
4 **Past Continuous** – Light Brown
5 **Present Perfect** – Green
6 **Futures** – Yellow
7 **Passives** – Orange
8 **Comparatives / Superlatives** – Red

I paint lots of pictures every year.

I'm painting my cat at the moment.

I've painted hundreds of pictures.

Tomorrow *I'm going* to paint my dog.

While *I was painting* the picture it started to rain.

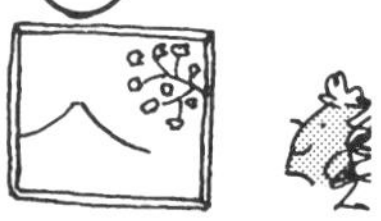

Last year *I painted* this picture in Japan.

It was painted by Leonardo in 1503.

I'll never paint as well as Picasso!

© English Experience, 25 Julian Road, Folkestone, Kent CT19 5HW

TEACHING NOTES

Unit 1 – Overview

Side A   7.5.11.1.9.2.10.4.6.8.3

Side B

Exercise 1
Because he sold all his fish. It's sold at the harbour. No. He hasn't.
He's selling fish to a lady. No. He has been selling fish since he was a boy.
Because they are going to buy lots of fish. It was sold on the beach.
The little ones are cheaper but the big one is the most delicious.
If he sells the salmon he'll buy her a present. He sells fish.
He was selling shellfish.

Exercise 2
Since I was a boy. Yes. In the old days fish was sold on the beach but now it's sold in the market. Yes. Yesterday one stole my hat while I was selling shellfish. I sold a lot yesterday but I haven't sold much today. I'll buy a surprise present for my wife. How about a nice salmon?

BRAIN-*friendly* tip:
Introduce 'Colour family' idea with coloured pens / pencils - enough for students to share.

Unit 2   'What's just happened?'
Exercise 1.
1. Yes. They have. 2. Yes. She has.
3. No. She hasn't. She has been to the swimming pool.
4. Yes they have. 5. No. He hasn't. 6. Yes. They have.
7. Yes. She has. 8. Yes. It has.

Exercise 2.
1. They've broken a window.
2. She's fallen off her skateboard.
3. She's been to the swimming pool. 4. They've robbed the bank. 5. He's caught three.
6. No. They're not. 7. No. She's just woken up. 8. No. He's missed it. It's (just) gone.

Exercise 3.
Where has Gwen been? How many fish has Gordon caught?
What have Greg and Grabber stolen? Is Gordon in time for the bus.

BRAIN-*friendly* tip:
(Switch on light / Drop book etc) Wear something Green as you move around the class doing things and prompt 'What have I just done? 'What's just happened?' questions.

Unit 3 – For / Since / Ago

Exercise 1.
1. They've been married for 25 years. They've lived there since 1992.
2. It arrived a few minutes ago/It's just arrived/It has already arrived.
The Zurich flight hasn't arrived yet.
3. No, because she's already got one/She bought one recently/
She's just bought one. She's had it since last summer.
4. It was built 600 years ago. He's had it since 1980.
He's been smoking for 15 years. 5. They've been there since
six pm/for nearly six hours. No. They've already eaten.
6. She started eight years ago/when she was six.
She's been there for eight years/since she was six. No. She's just got a new one.
She bought one recently. Only since two o'clock.

BRAIN-*friendly* tip:
Think of some 'marker words' (since / ago) as members of specific colour families (green / brown) while other words (for / recently) are 'visitors' to both.

Unit 4 – Past / Present Perfect

Exercise 1.
7.10.5.3
Examples
4. She walked 40 km yesterday.
6. He lived in Hong Kong from 1985-90.
8. She won Wimbledon a long time ago.
9. He has drunk two beers (so far).

Exercise 2.
5.8.9.4.2.1.10.6

BRAIN-*friendly* tip:
Use thick Dark Green and Brown pens to write master sentences for these pictures on big posters. Keep them as peripherals.

© English Experience, 25 Julian Road, Folkestone, Kent CT19 5HW

TEACHING NOTES

Unit 5 – Past Simple / Continuous
Examples:
2)
What were the ladies doing when it started to rain?
They were playing golf.
What did they do when it started to rain?
They went to the clubhouse for a drink.

3)
What was the waiter doing when the bulldog bit him?
He was carrying a tray of drinks.
What did he do when the dog bit him?
He dropped it and fell into the fountain.

4)
What was the nurse doing when the car crashed?
She was riding her bike.
What did she do when the car crashed?
She stopped and went back to help.

5)
What was the farmer doing when he found the vase?
He was digging in his field.
What did he do when he found the vase?
He sold it to a museum/antique dealér.

6)
What was the sentry doing when the officer arrived?
He was reading a girlie magazine.
What did he do when the officer arrived?
He stood to attention, saluted, and put out his cigarette.

7) What was she doing when the UFO landed?
She was picking flowers.
What did she do when it landed?
She went on board.

8)
What were they doing when the baby started to cry?
He was reading and she was drying her hair.
What did they do when the baby started to cry?
He(!) got up and changed its nappy.

> **BRAIN-*friendly* tip:**
> Show the Past Continuous as a long, wavy Light Brown line interrupted by a Dark Brown (Past) action:
> When I saw it (Dark Brown)
> I was digging. (Light Brown)

Unit 6 – Instructions
Exercise 1
plug switch take put Press Wait Take
put check empty throw/pour take

Exercise 2
read switch put in/insert choose
move click try check Ask

Exercise 3
Check prepare Put add cook Add Cook put
Bring Stir Heat Put Arrange Add Add Add/Arrange
add pour Sprinkle Put cook

> **BRAIN-*friendly* tip:**
> Collect examples of instructions in English from commercial products. Ask students to check how many of our Unit 6 verbs are included in these 'real-life' examples. What other verbs are used? Are they generally useful - or very specialised?

© English Experience, 25 Julian Road, Folkestone, Kent CT19 5HW

TEACHING NOTES

Unit 7 – Much / Many / A lot of

Exercise 1
a) a lot of. many. any
b) a lot of. weren't. aren't any
c) is a lot of. isn't much. isn't any
d) a lot of. much time. isn't any time
e) a lot of money. have much. haven't got any
f) a lot of. weren't many. there aren't any

> **BRAIN-*friendly* tip:**
> Ask the class working in 2's or 3's to brainstorm and, if possible, demonstrate examples of comparatives and superlatives.

Exercise 2
a) It'll swim to the island because there are a lot of people there - and there aren't many in the boat.
b) In 1930 there were a lot of trees and now there aren't any.
c) No. There isn't much water in it.
d) Because there's a lot of time before the next train.
e) He's foolish. When he has a lot of money, he spends it all very quickly - and then he doesn't have any money for food!
f) Because there aren't many (left).

Unit 8 – Comparatives and Superlatives

Exercise 1
For example:
Roger's house is smaller than Penny's.
Penny's house is more modern than Roger's.
Her car is not as old as his.
His camera takes better pictures than hers.
His dog is friendlier than hers.

> **BRAIN-*friendly* tip:**
> Use edible items (like smarties) to demonstrate a lot / not many / aren't any.
> Do the same with a liquid for not much / isn't any.

Exercise 2
for example:
Penny is the youngest.
Clarissa is the most elegant.
Roger has the most expensive camera.
Sam is carrying the heaviest bag.

Unit 9 – First Conditional

Exercise 1
How long will it take if she goes by bike?
How much will it cost if she goes by boat?
How much will it cost if she goes by car?
How long will it take if she goes by car?
And how long will it take her if she walks?
What will it cost if she hitchhikes?

> **BRAIN-*friendly* tip:**
> Use a dark blue pen and a yellow one to show how the 1st conditional sentence is constructed of 2 parts. If + Dark Blue (Present Simple) + Yellow (Future 'will')
> If she goes (Dark Blue) by plane it will (Yellow) take 1 hour.

Exercise 2
For example:
If she goes by car she'll need a map and lots of petrol.
If she goes by bike she'll need a new puncture repair kit and a pump.
If she hitchhikes she'll need a big sign, food - and a book on self defence.

TEACHING NOTES

Unit 10 – Present and Past Passive

a) The castle was built in 1157. It was attacked and burned in 1304. In 1992 the ruins were bought for £1m. The castle was rebuilt and was opened as a hotel in 1998. It is used for Conferences and Trade Fairs.
b) The cotton for Jazy shirts is grown in India. It is taken to cotton mills and made into cloth. The cloth is taken to the factory where it is made into shirts. It is sold in Jazy shops all over Europe and America.
c) Last Friday Kanahz was hit by an earthquake. Some people were rescued but over 300 were killed and thousands were injured. Almost all the buildings were damaged. Aid (food, medicine, clothing) was sent to Kanahz by helicopter and truck.
d) The picture was painted by Leonardo in 1502. In 1984 it was sold at auction for £84m. It was bought by the Tokyo Museum of Art and was put on display there. In 1990 it was stolen but yesterday it was found again. Unfortunately it is badly damaged.

Exercise 2
When was the picture painted? When was it stolen? Was it damaged?

Exercise 3
Where is the cotton for JAZY shirts grown?
And where are the shirts made? How are they sold?

BRAIN-*friendly* tip:
Use an orange pen, a dark blue pen, and a brown pen to show that all passives have an 'orange bit' as their second part. Present-Passive: It is (Blue) taken (Orange). Past Passive: It was (Brown) taken (Orange).

Unit 11 – Describing things

a) What's it called? When was it built? What's it made of? How high is it?
b) How far is it? How long does it take?
c) What's in the jar? How big is it? How heavy is it? What size jar is it? What's on the label?
d) What type/make of car is it? What colour is it? How fast is it? What's its top speed? How much does it cost? What's the price? When was it made? How old is it?
e) How long is the pipe? What's the diameter? What's it made of? Where was it made?
f) How long is the pool? How deep is it? What's the temperature?
g) What's this thing called? What's it used for?

BRAIN-*friendly* tip:
Play 20 Questions - and also have a bag of 'surprise items' for question making practice (even better if students contribute some of the items). Give help so that people don't get frustrated if they can't deduce the object.

Unit 12 – Revision

Exercise 1
10.2.11.1.9.4.3.6.8.7.5

Exercise 2
I've been flying for three years. If I fly for another two years I'll become a senior stewardess. I fly for Brain-friendly Airways. Not many planes fly as many passengers as this one. "Put your chair in the upright position and fasten your seat belt." I've flown to a lot of interesting places. Yes. When we were flying to Sydney a passenger proposed to me. No. All our pilots are very experienced. Yes. Next month for example I'm going to fly on the New York route. We're flying over the Andes.

The correct colours are: a-orange, b-dark green, c-dark blue + yellow, d - dark blue, e - orange, f - light green, g - light brown + dark brown, h - red, i - light blue, j - dark blue, k - yellow

BRAIN-*friendly* tip:
Have music playing (gently) in the background as students work on their grammar. Refresh your own memory of the colour code system in the Teachers' Notes.

© English Experience, 25 Julian Road, Folkestone, Kent CT19 5HW

PICTURES OF ENGLISH TENSES LEVEL 2 — 1A

How quickly can you connect the pictures to the right sentences?

Picture
- ☐ Here come some tourists. *I'm going to sell* lots of fish.
- ☐ *I've been selling* fish ever since I was a small boy.
- ☐ Most of the fish *is sold* in the market.
- ☒ **I *sell* fish at the harbour.**
- ☐ These are *cheaper* but this one is the *most delicious*.
- ☐ *I'm selling* her some fish for dinner.
- ☐ In the old days fish *was sold* on the beach.
- ☐ Yesterday I sold *a lot* but I haven't sold *much* today.
- ☐ Recently while I was selling shellfish a seagull *stole* my hat.
- ☐ If I *sell* this beauty *I'll buy* my wife a present.
- ☐ Yesterday I *sold* everything to a restaurant owner.

Hello. I'm Salty Sam. *I sell fish at the harbour.*

The "BIGGER PICTURE" – 1A

Mark Fletcher

© English Experience, 25 Julian Road, Folkestone, Kent CT19 5HW

1B PICTURES OF ENGLISH TENSES LEVEL 2

The 'BIGGER PICTURE'

Exercise 1

Use the Picture page 1A to answer these questions.

Why was Sam very happy yesterday?
....

Where is most of the fish sold?
....

Has Sam sold much fish today?
....

What's he doing?
....

Is selling fish a new job for Sam?
....

Why is Sam looking happy?
....

How was fish sold in the old days?
....

What is the difference between the little fish and the big one?
....

Perhaps Sam's wife will have a nice surprise. Why?
....

Sam's in the pub,– but what does he do most of the week?
....

What was Sam doing when the seagull took his hat?
....

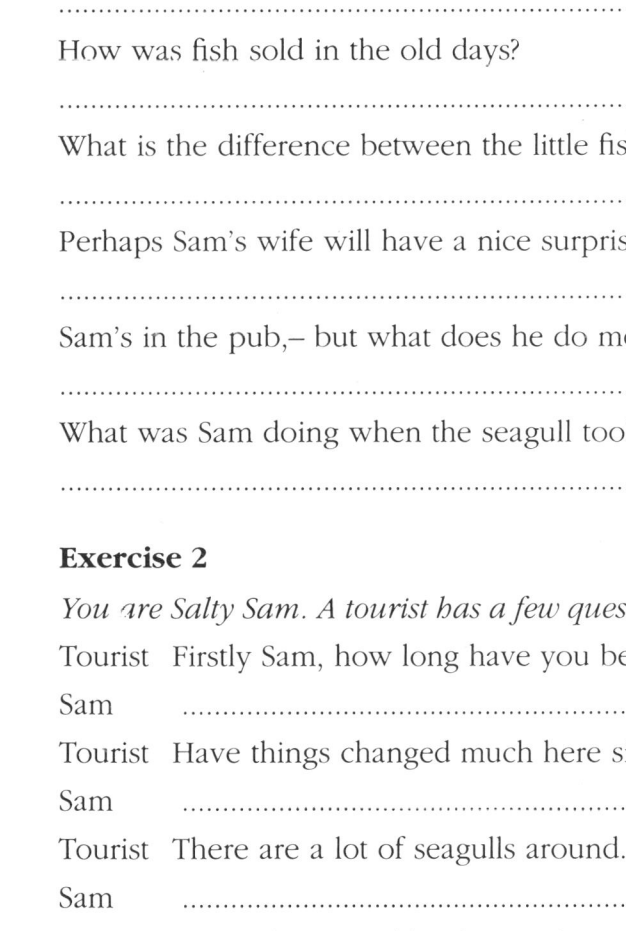

Exercise 2

You are Salty Sam. A tourist has a few questions for you.

Tourist Firstly Sam, how long have you been selling fish?
Sam   ....
Tourist Have things changed much here since the old days?
Sam   ....
Tourist There are a lot of seagulls around. Are they a problem?
Sam   ....
Tourist What's business like this week?
Sam   ....
Tourist That's a magnificent salmon - it's worth a lot of money.
         What will you do if you sell it?
Sam   .... 
Tourist And finally, I'd like to buy some fish. What do you recommend?
Sam   ....

© English Experience, 25 Julian Road, Folkestone, Kent CT19 5HW

PICTURES OF ENGLISH TENSES LEVEL 2   2A

Gaby Green says....

"What's just happened?"

'I've just ....' – 2A

Mark Fletcher

© English Experience, 25 Julian Road, Folkestone, Kent CT19 5HW

2B PICTURES OF ENGLISH TENSES LEVEL 2

'I've just...'
What's happened to the Green family?

Exercise 1

Picture			
1	Have the **Green** boys broken the window?	Yes. They have.
2	Has Gaby **Green** fallen off her skateboard?	....
3	Has Gwen **Green** been to the shops?	....
4	Have Greg and Grabber **Green** robbed the bank?	....
5	Has Gordon **Green** caught ten fish?	....
6	Have George and Gaby **Green** finished dinner?	....
7	Has Greta **Green** just woken up?	....
8	Has the green bus gone?	....

Exercise 2

Use the Picture page 2A to answer these questions.

Picture			
1	What have the boys done?	They've broken a window.
2	What's happened to Gaby?	....
3	Where has Gwen been?	....
4	What have Greg and Grabber done?	....
5	How many fish has Gordon caught?	....
6	Are George and Gaby going to start dinner?	....
7	Is Greta asleep?	....
8	Is Gordon in time for the green bus?	....

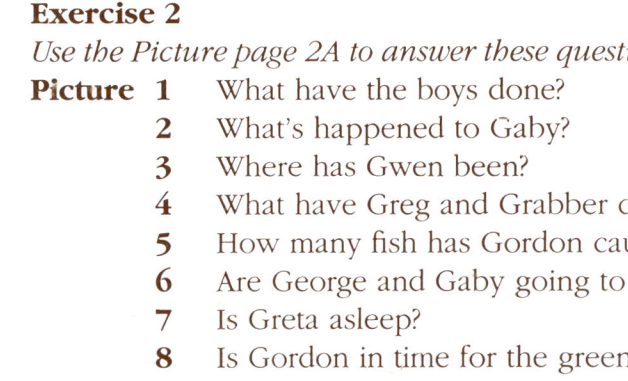

Exercise 3

You write the questions.

Q ....
A She's been to the swimming pool.

Q ....
A He's caught three.

Q ....
A They've stolen some money.

Q ....
A Yes – and now they must do the washing up.

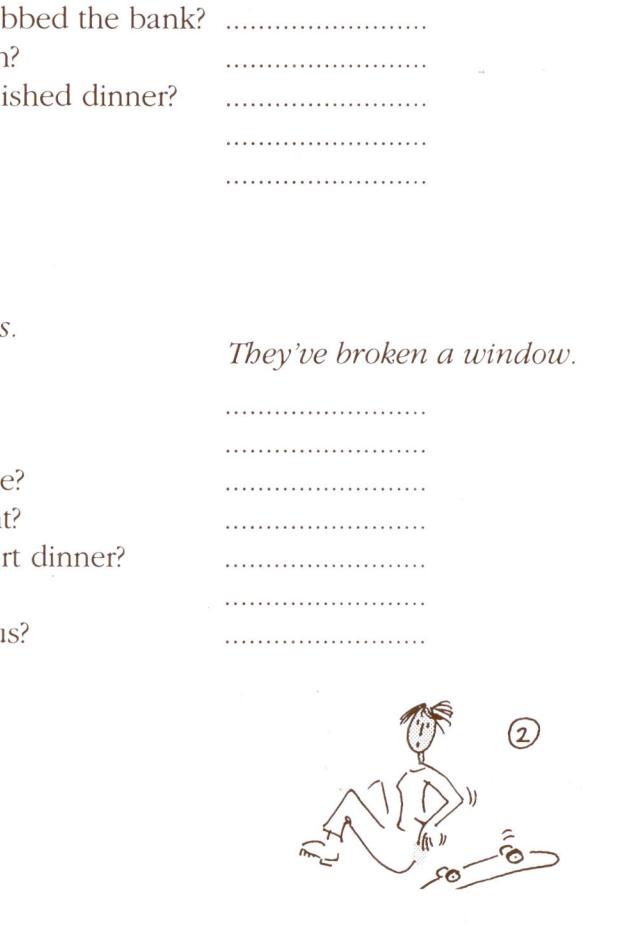

Write the correct sentences under these two pictures.
*Underline the sentences in **Green**.*

© English Experience, 25 Julian Road, Folkestone, Kent CT19 5HW

PICTURES OF ENGLISH TENSES LEVEL 2   3A

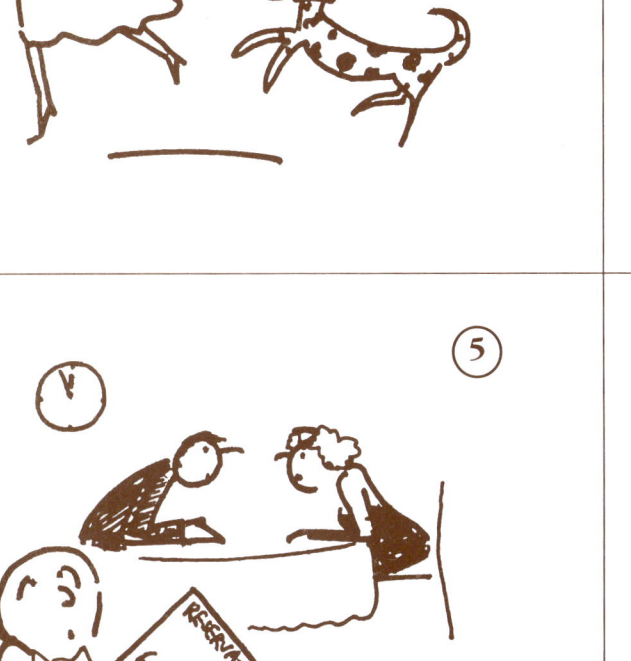

For / Since / Ago – 3A

3B PICTURES OF ENGLISH TENSES LEVEL 2

Use **For** or **Since** or **Ago**
*(or **Yet** or **Already** or **Just** or **Recently**)*

Exercise 1

1. How long have they been married? ....
   How long have they lived in this house? ....

2. I'm meeting someone on the Amsterdam flight.
   Is there a problem? ....
   Has the Zürich flight landed? ....

3. Is she's going to buy a new hat? ....
   How long has she had her dog? ....

4. Is the castle very old? ....
   Has he had the car long? ....
   When did he start smoking? ....

5. How long have they been in the restaurant? ....
   Are they about to have dinner? ....

6. When did she start at the school? ....
   How many years has she been there? ....
   Does she need a new bicycle? ....
   Has it been sunny all day? ....

For / Since / Ago / Yet / Already / Just / Recently

Some words belong to the **Brown** (Past) family.

                                                Others belong to the **Green** (Present Perfect) family.

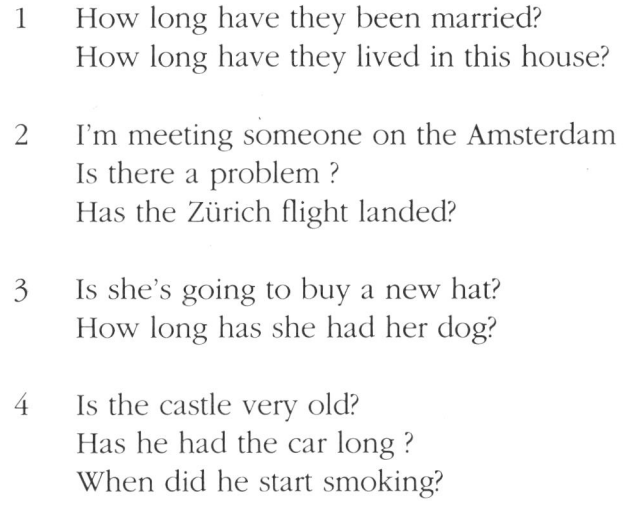

Brown family

Green family

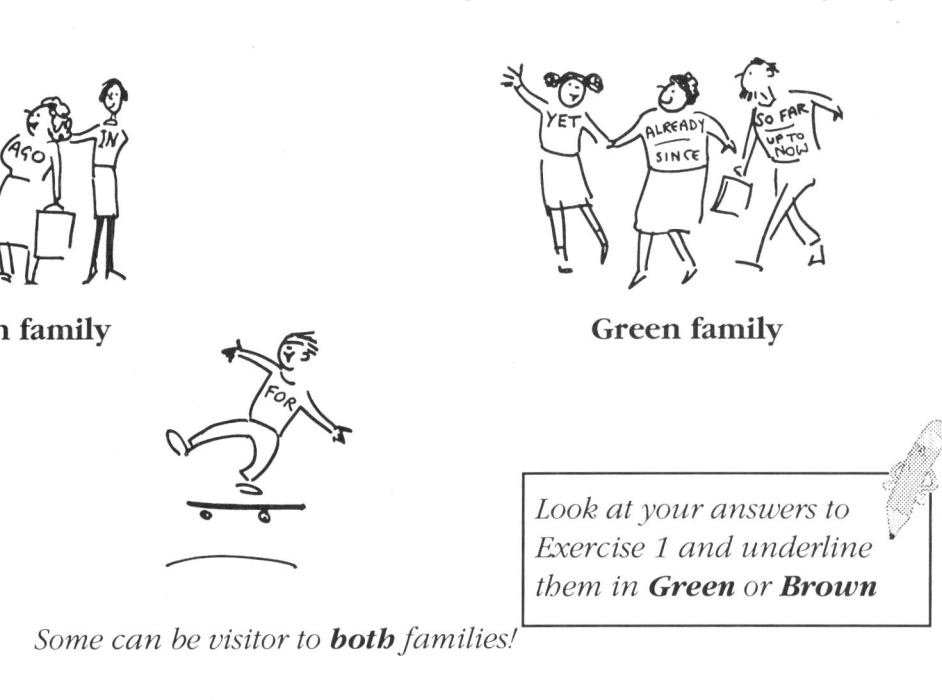

*Some can be visitor to **both** families!*

*Look at your answers to Exercise 1 and underline them in **Green** or **Brown***

© English Experience, 25 Julian Road, Folkestone, Kent CT19 5HW

PICTURES OF ENGLISH TENSES LEVEL 2 — 4A

Glenda Green says... **Betty Brown** says...

"I've worked here for three weeks."

"I worked there for twenty years"

'What has she done / What did she do' — 4A

4B PICTURES OF ENGLISH TENSES LEVEL 2

Glenda Green & Betty Brown

Glenda **Green** is twenty-two years old. She is enjoying her new job. Glenda works in a bank. She *has worked* there for three weeks.

Betty **Brown** is in town. She *worked* for the bank for twenty years. She retired in 1985.

Exercise 1

Write these sentences under the correct picture on the picture page.

a   She has won the tournament.
b   He drank too much.
c   He's lived in Hong Kong for a long time.
d   He's walked quite a long way.

And now write sentences under the other four pictures.

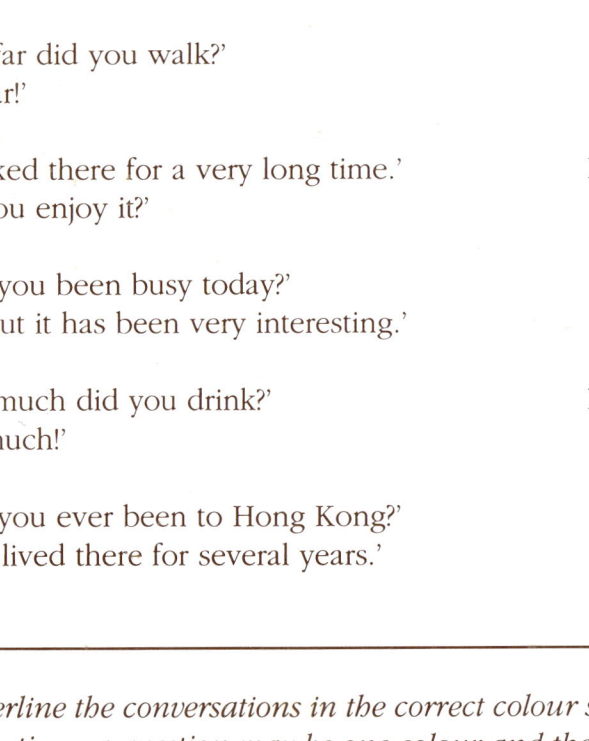

Exercise 2

Which pictures connect with these conversations?

'How long have you lived in Hong Kong?'   Picture ....
'Since 1985.'

'Have you ever won a tennis championship?'   Picture ....
'Yes. I won Wimbledon when I was a teenager.'

'How much have you drunk?'   Picture ....
'I don't know. Ask the barmaid.'

'How far did you walk?'   Picture ....
'Too far!'

'I worked there for a very long time.'   Picture ....
'Did you enjoy it?'

'Have you been busy today?'   Picture ....
'Yes, but it has been very interesting.'

'How much did you drink?'   Picture ....
'Too much!'

'Have you ever been to Hong Kong?'   Picture ....
'Yes. I lived there for several years.'

---

*Underline the conversations in the correct colours – **Green** or **Brown**.
(Sometimes a question may be one colour and the answer a different colour!)*

© English Experience, 25 Julian Road, Folkestone, Kent CT19 5HW

PICTURES OF ENGLISH TENSES LEVEL 2   5A

Past / Past Continuous – 5A

5B PICTURES OF ENGLISH TENSES LEVEL 2

Past / Past Continuous

Picture Vocabulary - you will need these words to talk about the pictures

1. a thief   to rob (someone or a place)   to steal (something)   a safe   jewels
2. club house   3   bulldog   tray   to bite
4. a nurse   to crash   5   to dig   vase
6. sentry   sentry box   to salute   to stand to attention
7. to land   to pick   UFO/flying saucer   8   to cry

Exercise 1

Complete the conversations. For example:

1. What was the thief doing when the policeman arrived?

   He was stealing the jewels.

   What did he do when the policeman arrived?

   He jumped out of the window/ran away/hit the policeman.

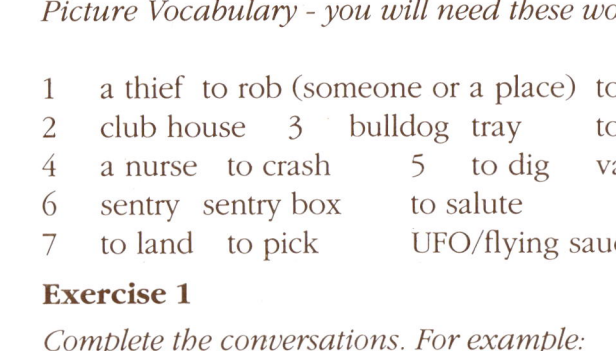

2. What were the ladies doing when it ....?

   They..

   What did they do ....?

   They ....

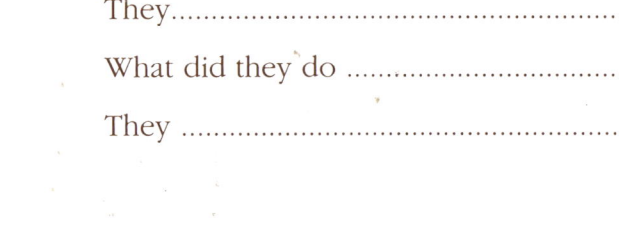

*Use **Dark Brown** to underline the Simple Past and **Light Brown** for the Past Continuous.*

Exercise 2

Write 4-line conversations for the other six pictures.

3 ....   6 ....

4 ....   7 ....

5 ....   8 ....

PICTURES OF ENGLISH TENSES LEVEL 2 — 6A

Using your new Toaster

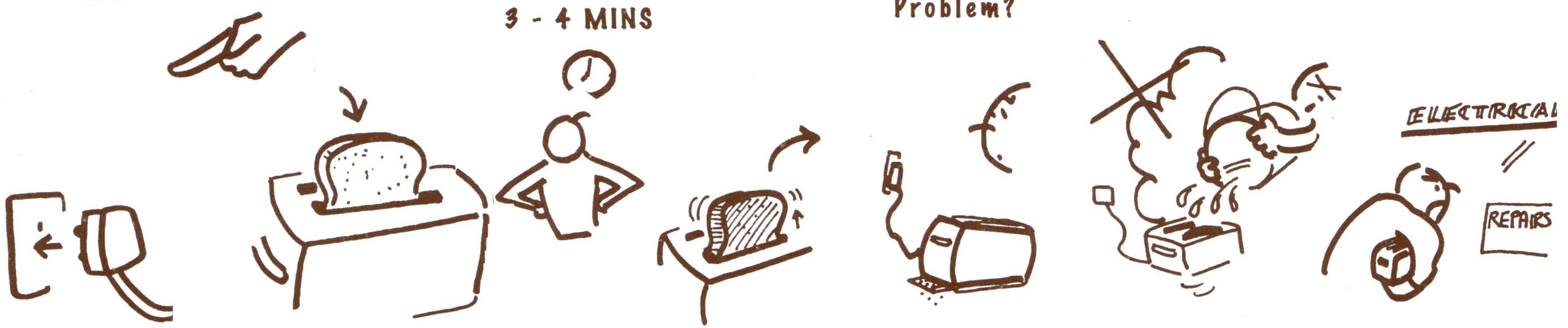

Loading a Computer Game

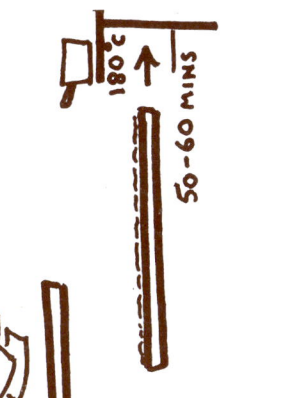

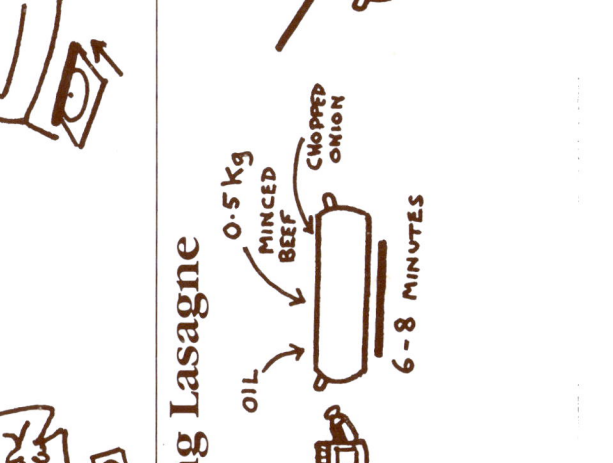

Making Lasagne

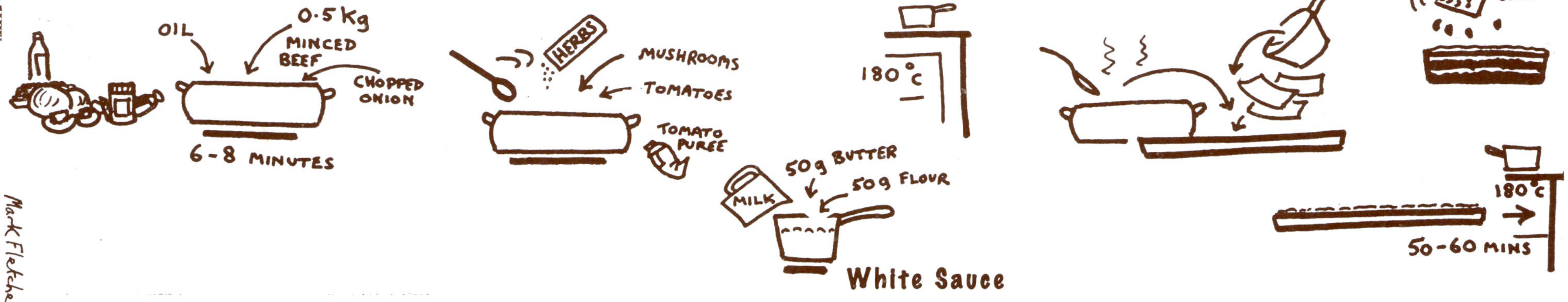

Instructions – 6A

© English Experience, 25 Julian Road, Folkestone, Kent CT19 5HW

6B   PICTURES OF ENGLISH TENSES LEVEL 2

Instructions

Use these verbs to complete the three instructions. You will need some of the verbs more than once.

check	connect	plug	prepare	switch	take	insert	put	add	choose
press	move	wait	cook	click	bring	empty	stir	try	heat	place/arrange
throw	pour	sprinkle	read	turn	ask

Exercise 1 – Using your new Toaster

To make the toaster work, ... it in and .... it on. Cut a piece of of bread (or .... a slice from the packet) and .... it into the toaster. .... down the 'toaster bar'. .... a few minutes until the toast pops up. .... it out and .... it on a plate. If there is a problem and the toaster doesn't work, ... that it properly plugged in and switched on. You may need to .... the crumb tray. *Warning:* Under no circumstances .... water on the toaster! If there is a major problem, ... it to an electrical repair shop.

Exercise 2 – Loading a Computer Game

Before you start, ... the instructions!
.... on the computer and .... in / .... the C.D.Rom. When you see the menu on the screen, ... the game you want. (........ the cursor on the screen with the mouse and when it is in the right place, ... on the mouse.) If the progamme doesn't run first time, then .... again. If it still doesn't work .... that the mouse is properly connected to the keyboard and that the keyboard is properly connected to the computer.
It still doesn't work? .... for help!

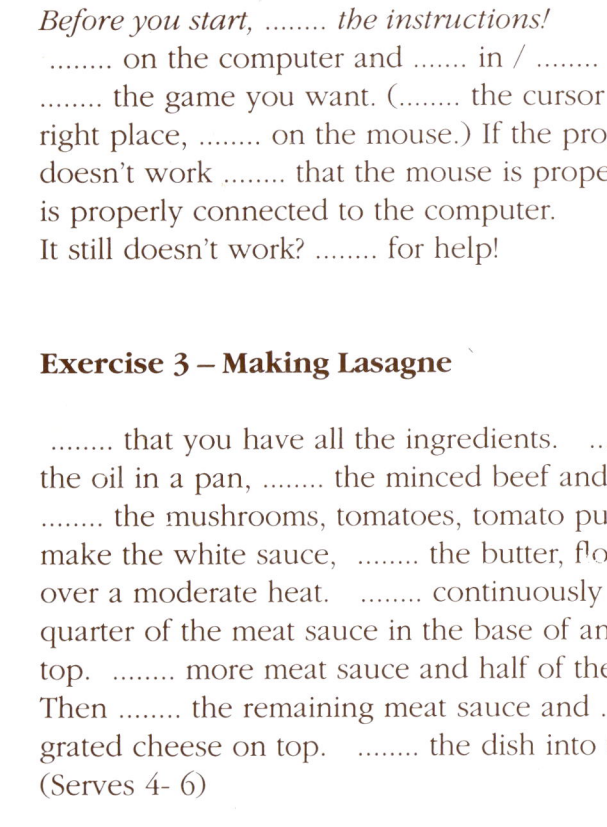

Exercise 3 – Making Lasagne

 .... that you have all the ingredients.   .... the vegetables (wash them and cut them up). .... the oil in a pan, ... the minced beef and onion and .... for 6-8 minutes stirring occasionally. .... the mushrooms, tomatoes, tomato puree, herbs and seasoning. .... for ten minutes. To make the white sauce, ... the butter, flour and milk into a saucepan. .... gently to the boil over a moderate heat.   .... continuously until smooth.   .... the oven to 180 Deg C.   .... a quarter of the meat sauce in the base of an ovenproof dish. .... half of the lasagne sheets on top. .... more meat sauce and half of the white sauce.   .... the remaining lasagne sheets. Then .... the remaining meat sauce and .... the rest of the white sauce over it. .... the grated cheese on top. .... the dish into the preheated oven and .... for 50-60 minutes.
(Serves 4- 6)

Exercise 4

Look at the picture instructions again (Side A).
Explain one of the tasks to your neighbour.
Use 'First', 'After that', 'Next', 'You should...', 'Remember to ...'.

*Choose the description you like best and underline all the instruction verbs in **Dark Blue**.*

© English Experience, 25 Julian Road, Folkestone, Kent CT19 5HW

PICTURES OF ENGLISH TENSES LEVEL 2   7A

A lot of	much (or many)	any

Much / Many / A lot – 7A

© English Experience, 25 Julian Road, Folkestone, Kent CT19 5HW

7B   PICTURES OF ENGLISH TENSES LEVEL 2

Much / Many / A lot of

Exercise 1

Look at the picture page 7A and make sentences.

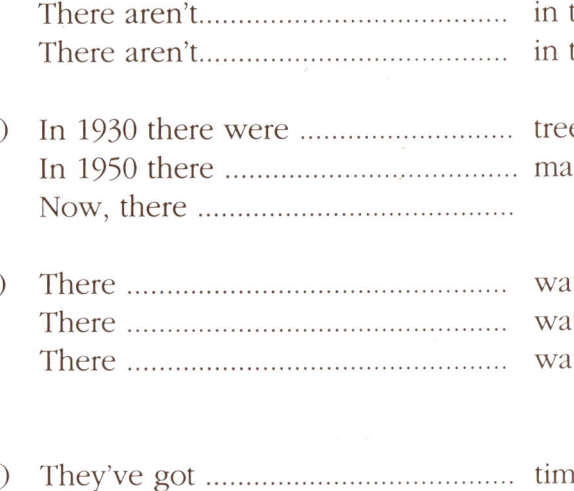

a) There are .... people on the island.
   There aren't.................................. in the boat.
   There aren't.................................. in the sea.

b) In 1930 there were .... trees here.
   In 1950 there .... many.
   Now, there ....

c) There .... water in the reservoir.
   There .... water in the bucket.
   There .... water in the bottle.

d) They've got .... time before the next train.
   She hasn't got .... before the shop closes.
   There .... to deliver the message.

e) On pay day I had ....
   Next day I didn't .... money.
   Today I ....

f) There were .... cakes in the shop at 3 pm.
   By 4 pm there ....
   And now ....

Exercise 2

Use the pictures to answer the questions.

a) The sea monster is very hungry - and its favourite food is people! Will it swim to the island or to the boat - and why?
   ....

b) What's the difference between 1930 and now?
   ....

c) Is the bucket full now?
   ....

d) The travellers at the station have decided to go for a walk. Why?
   ....

e) What is the young man's life style? Does he use his money wisely?
   ....

f) You want to buy cakes. It's 4pm. You must hurry. Why?
   ....

*Look at your answers to Exercise 2. Underline Much / Many / Any / A lot of in **Red**.*

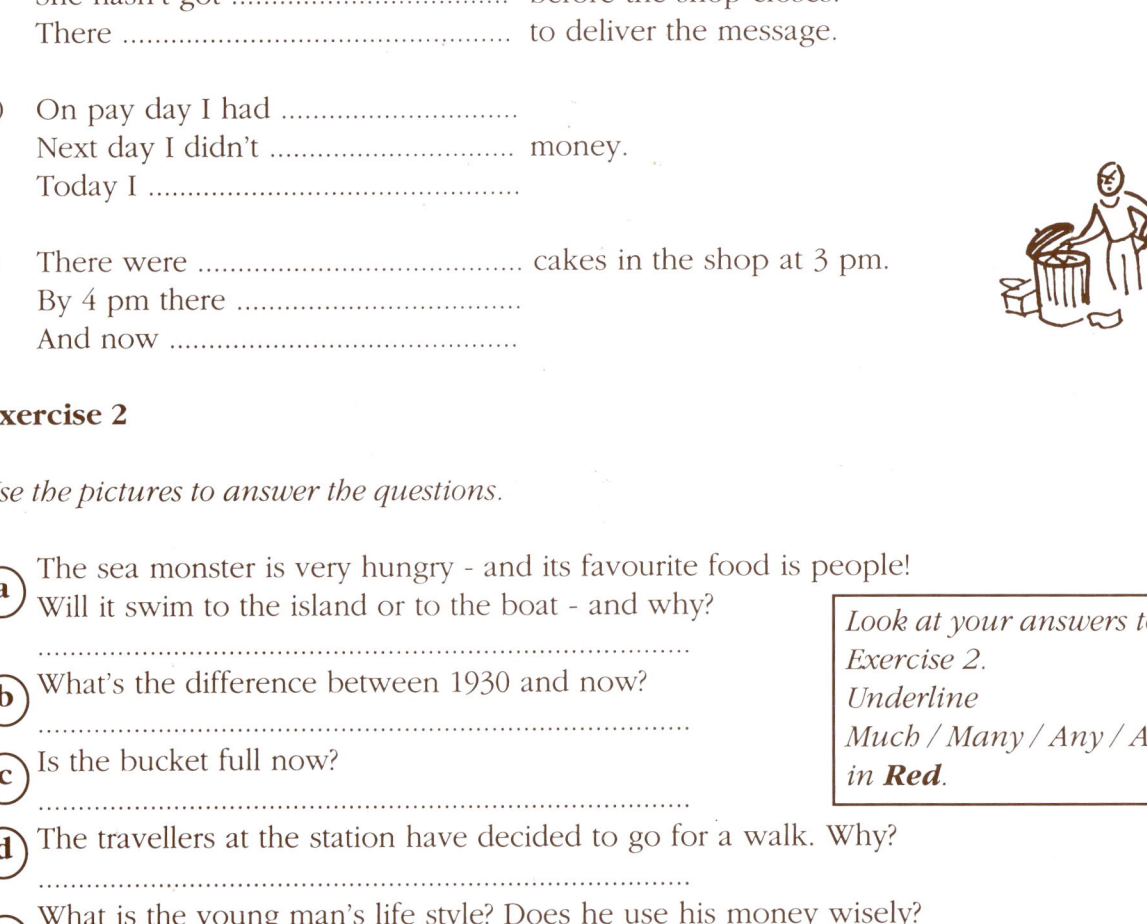

© English Experience, 25 Julian Road, Folkestone, Kent CT19 5HW

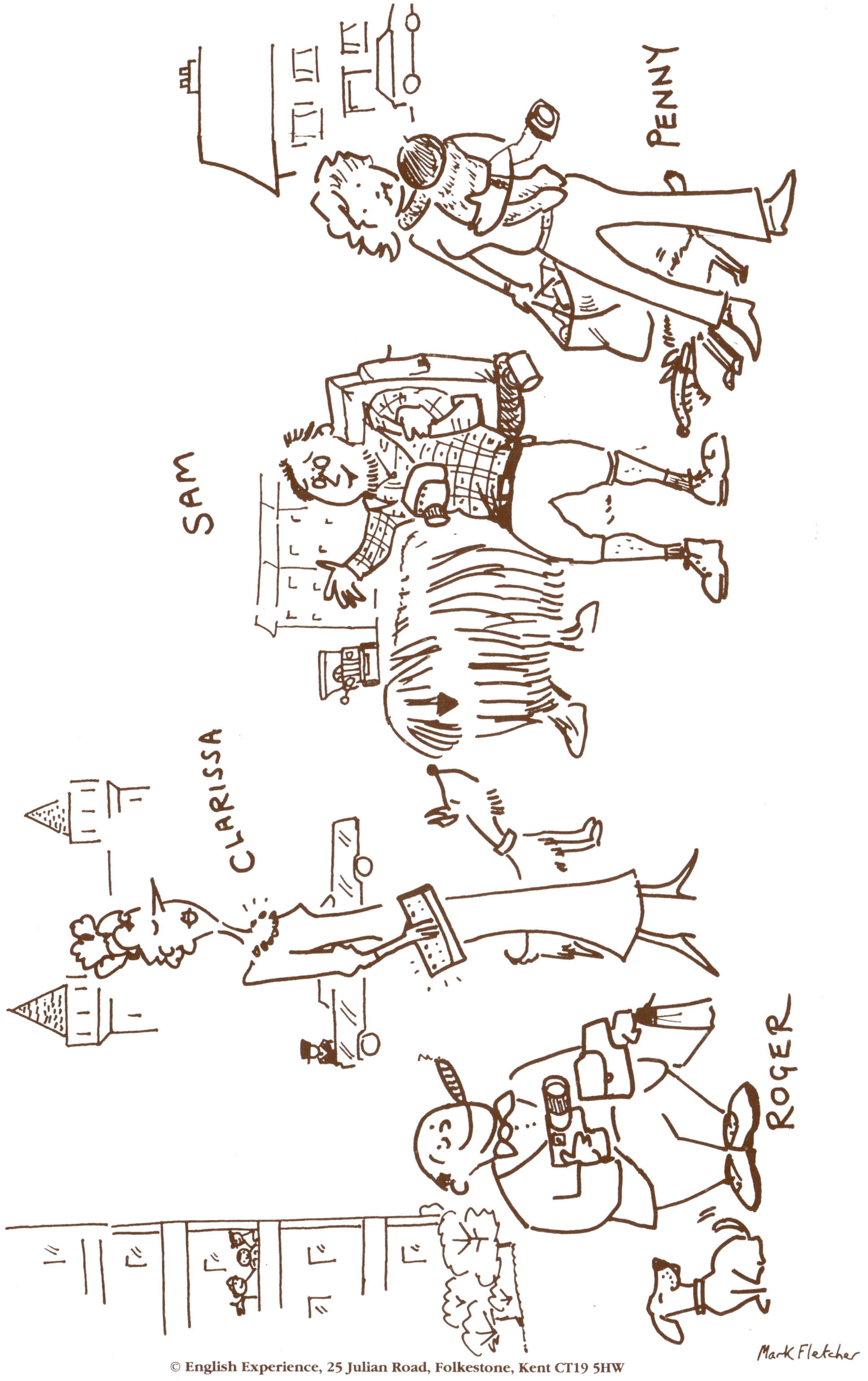

PICTURES OF ENGLISH TENSES LEVEL 2 — 8A

Comparatives and Superlatives – 8A

8B PICTURES OF ENGLISH TENSES LEVEL 2

Comparatives and Superlatives

Exercise 1

Let's compare two people, Roger and Clarissa.

Roger is **shorter than** Clarissa.

Her nose is **longer than** Roger's (and more pointed).

Her home is **more expensive than** his.

Roger's dog isn't **as big as** Clarissa's.

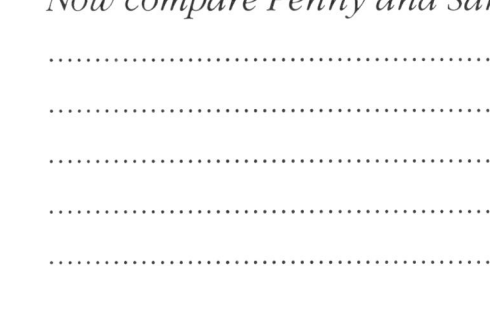

Now compare Penny and Sam.   Write 5 things about them.

....
....
....
....
....

Exercise 2

*Looking at all **four** people. Is it true? false? not sure? that . .*

Roger has **the most** children.   Sam lives in **the oldest** house.

Penny has **the most intelligent** dog.

Write 6 more things about them.

....
....
....
....
....
....

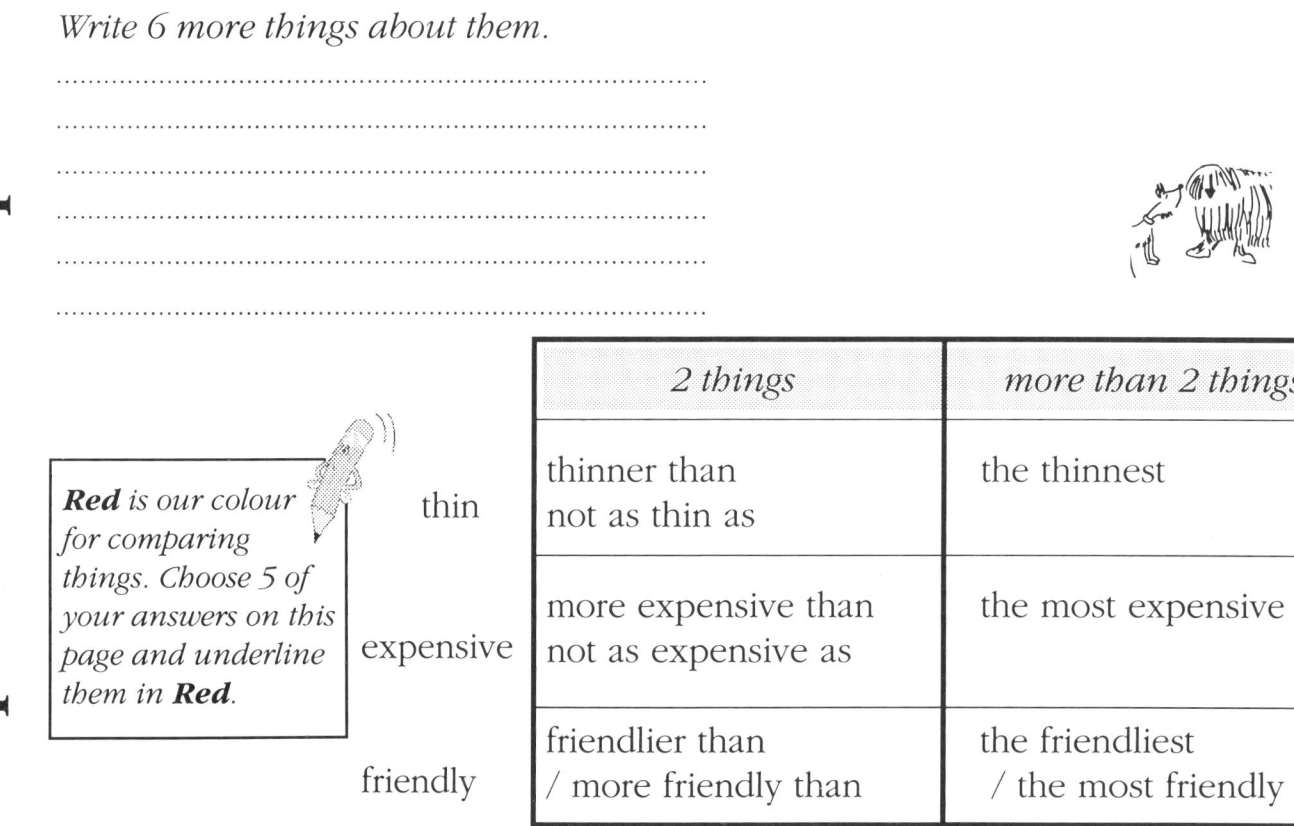

Red *is our colour for comparing things. Choose 5 of your answers on this page and underline them in **Red**.*

	2 things	more than 2 things
thin	thinner than / not as thin as	the thinnest
expensive	more expensive than / not as expensive as	the most expensive
friendly	friendlier than / more friendly than	the friendliest / the most friendly

© English Experience, 25 Julian Road, Folkestone, Kent CT19 5HW

PICTURES OF ENGLISH TENSES LEVEL 2 — 9A

Granny wants to go to Scotland for a holiday . . .

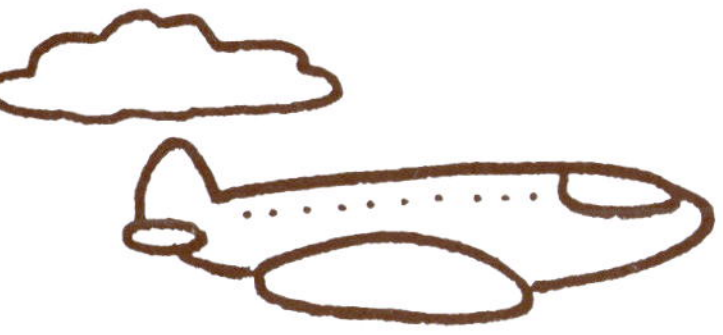

There are many different ways she can go there ...

✈	1 Hour	£80
 🚢	3 Days	£140
 🚗	12 Hours	£40
 🚲	5 Weeks	£17
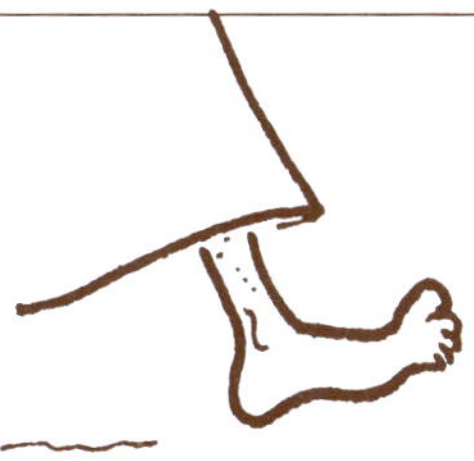 🦶	6 Months	2 Pairs of New Shoes
 👍	2-3 Days	£0

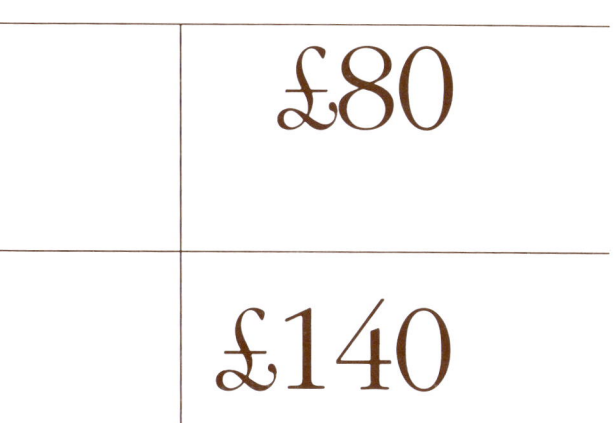

1st Conditional – 9A

© English Experience, 25 Julian Road, Folkestone, Kent CT19 5HW

Mark Fletcher

9B PICTURES OF ENGLISH TENSES LEVEL 2

1st Conditional

Exercise 1

Granny wants to go to Scotland for a holiday. There are many different ways she can go.
If she goes by plane **it will take** one hour and **it will cost** £80.

How much will it cost is she goes by boat?
How long will it take if she goes by boat?
You are talking to a Travel Agent. Complete the conversation.

You	....
Agent	It'll take five weeks I think.
You	....
Agent	It'll cost £140.
You	....
Agent	About £40 for petrol.
You	....
Agent	If she drives fast? It'll take about twelve hours.
You	....
Agent	It's a long way... I think it'll take her about six months.
You	....
Agent	It won't cost anything if she hitchhikes.

Exercise 2

She'll need an airline ticket if she goes by plane.
What will she need if she goes a different way?

If she goes by ....
....
....
....

*The 1st Conditional consists of an IF + Present (**Blue**) part, and a Future (**Yellow**) part.*
*Use **Blue** and **Yellow** to underline 5 good examples of this.*

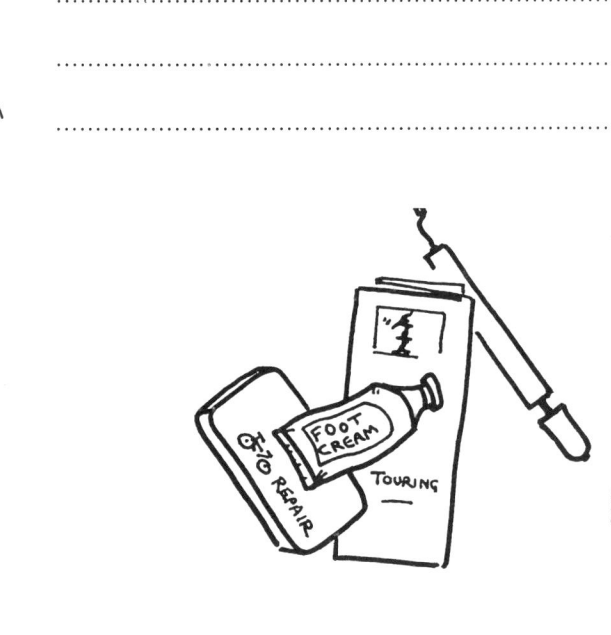

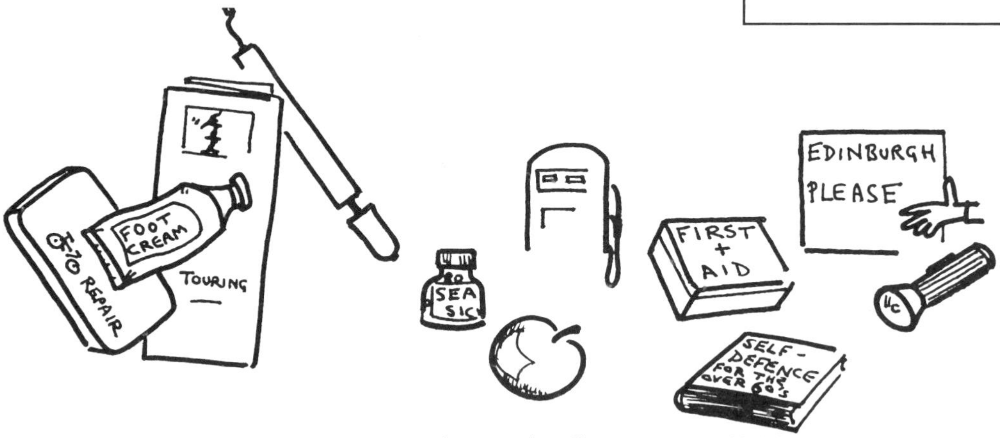

© English Experience, 25 Julian Road, Folkestone, Kent CT19 5HW

PICTURES OF ENGLISH TENSES LEVEL 2 — 10A

The 'Old Castle' Hotel
(a)
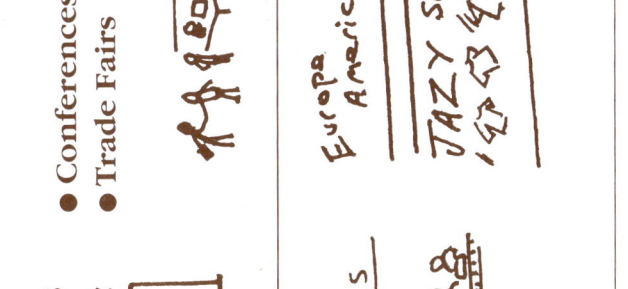

JAZY SHIRTS
(b)

EARTHQUAKE
(c)
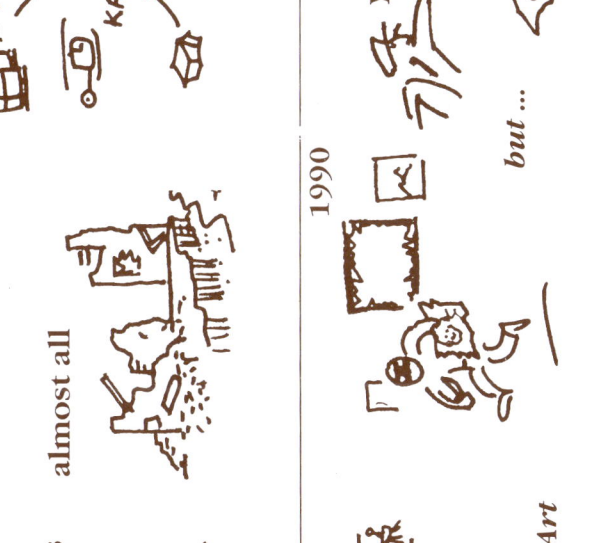

Lisa's Sister
(d)

Mark Fletcher

© English Experience, 25 Julian Road, Folkestone, Kent CT19 5HW

Present and Past Passive — 10A

10B PICTURES OF ENGLISH TENSES LEVEL 2

Present and Past Passive

Exercise 1
Use the following verbs to tell the four stories on side A in the Passive.
You will need some of the verbs more than once.

paint	hit	grow	build	take	sell	make (into)
buy	attack	make	kill	burn	rescue	find
send	damage	put	use (for)	steal	destroy	open

a) The castle **was built** in 1157. It .... and **was burned** in 1304.

In 1992 the ruins .... for £1m.

The castle .... and .... as an hotel in 1998.

These days it .... for Conferences and Trade Fairs.

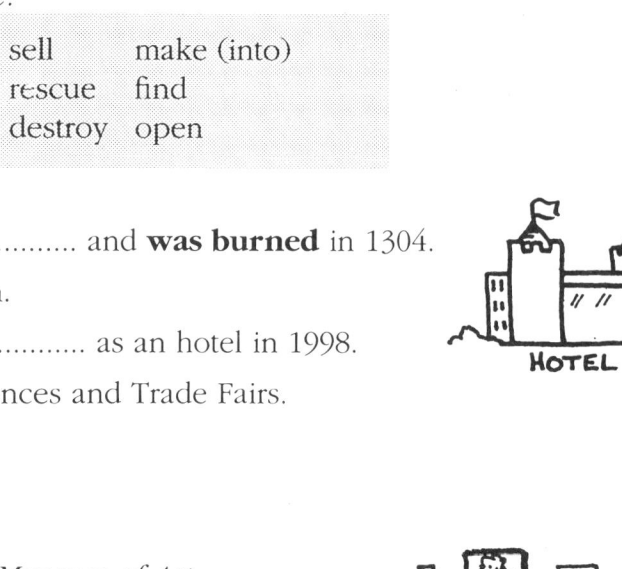

Now continue with the later stories.

Exercise 2
You are interviewing the Director of the Tokyo Museum of Art.

You   ....
Director   In 1502 - by Leonardo da Vinci.
You   ....
Director   One night in 1990.
You   ....
Director   It was damaged when it was cut out of the frame.
            I'm not sure if it can be repaired.

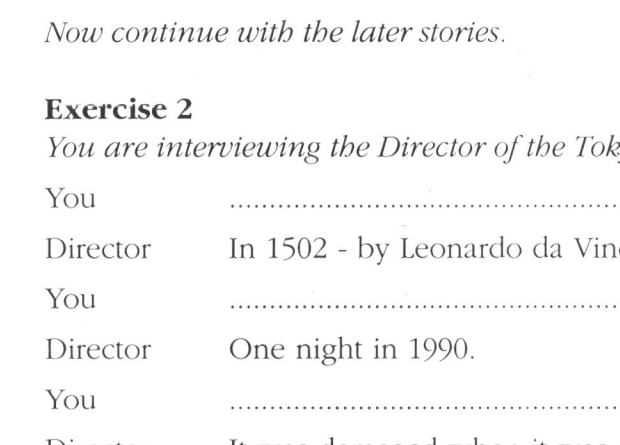

Exercise 3
You are interviewing the owner of the JAZY shirt company.

You   ....
Director   The cotton we use is grown in India.
You   ....
Director   The shirts are made in our textile factories.
You   ....
Director   They are sold in our retail branches
            - over a thousand of them - all over the world.

Exercise 4
You are a survivor of the earthquake at Kanahz. Describe what happened.

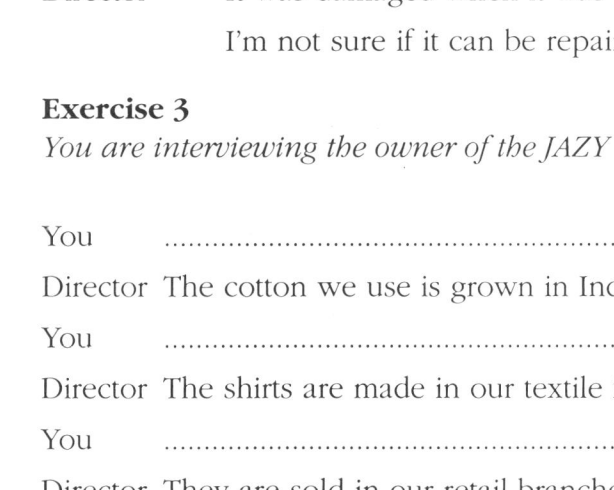

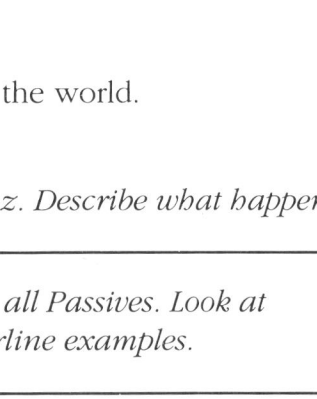

Orange is our colour for all Passives. Look at Exercises 1 - 3 and underline examples.

© English Experience, 25 Julian Road, Folkestone, Kent CT19 5HW

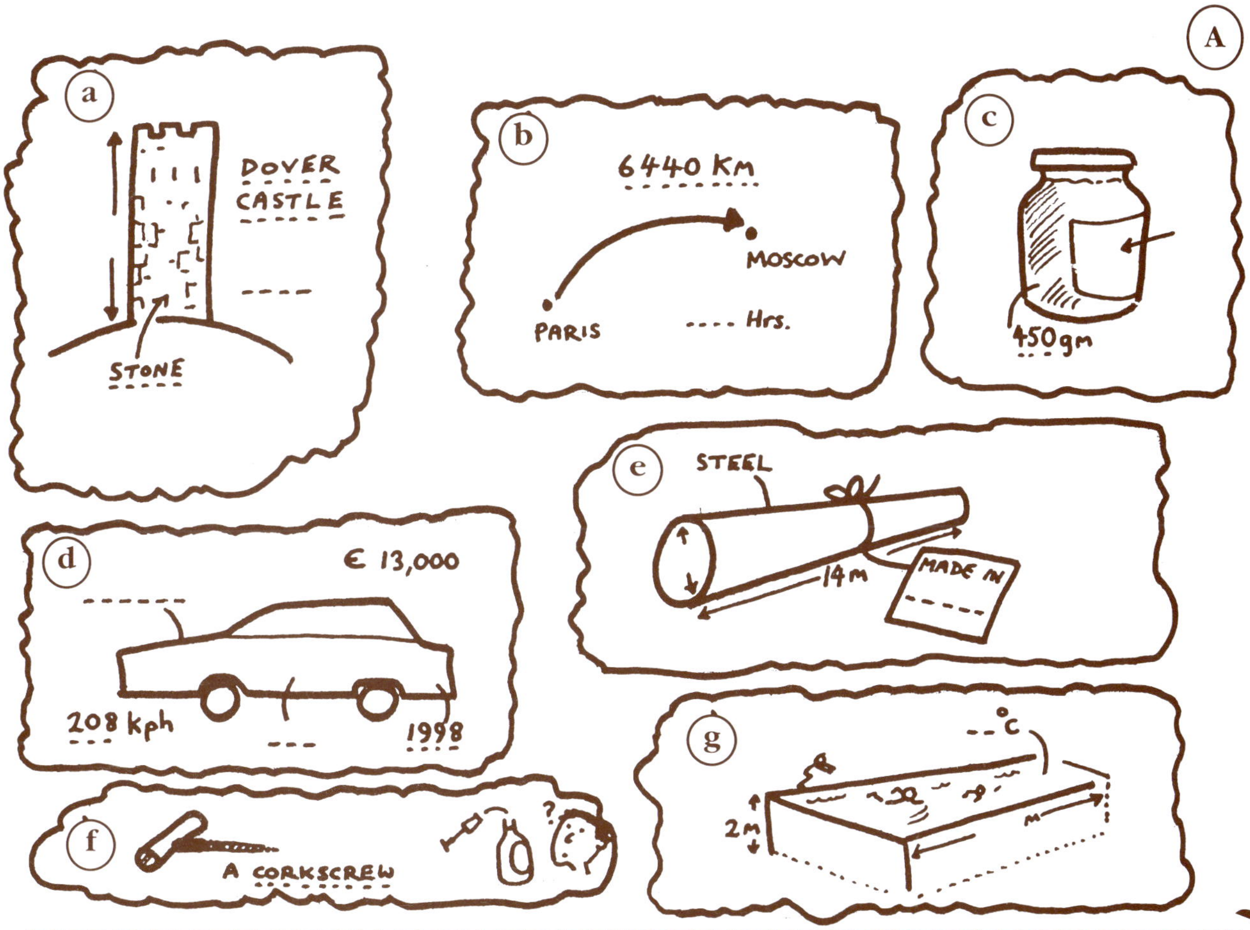

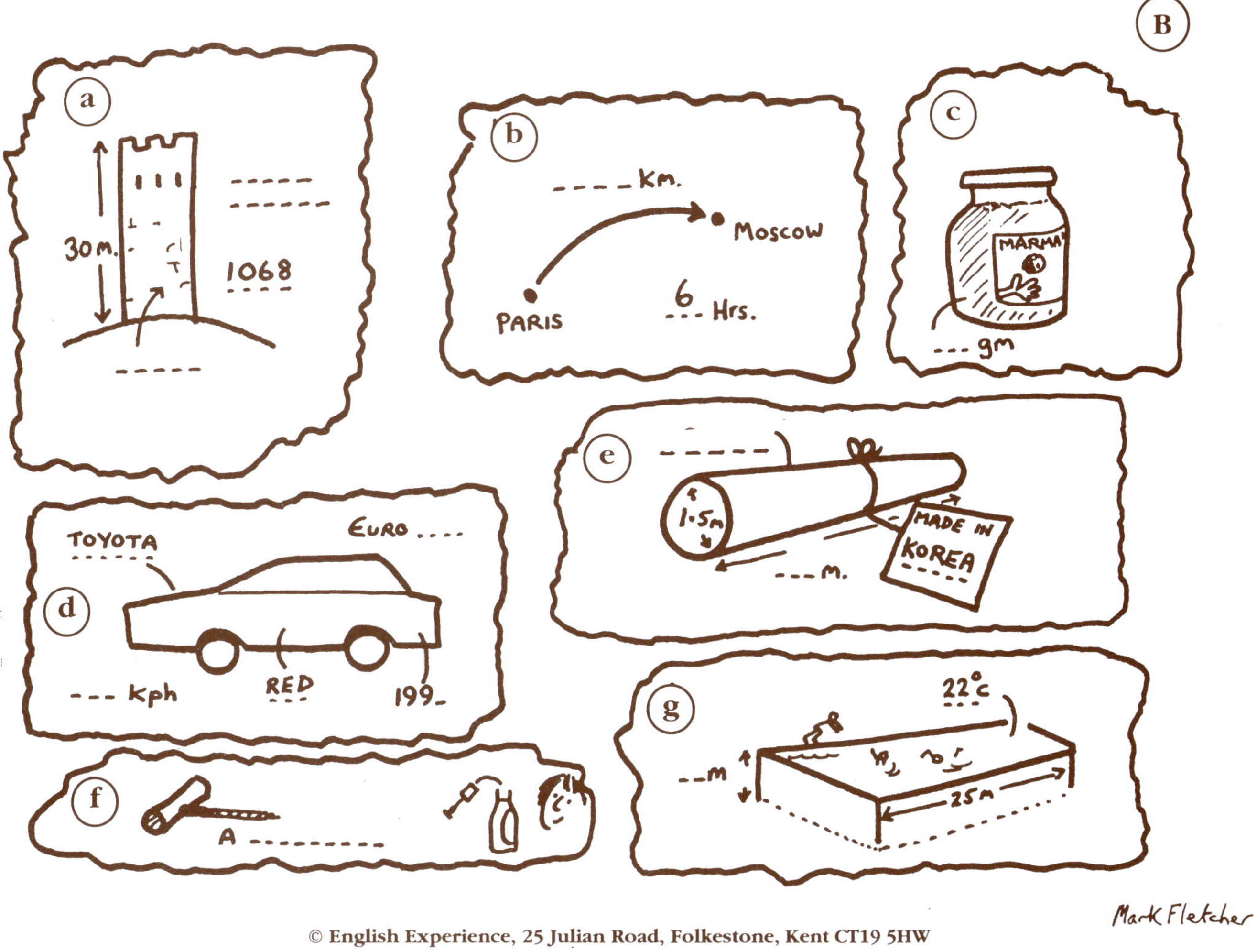

PICTURES OF ENGLISH TENSES LEVEL 2   11A

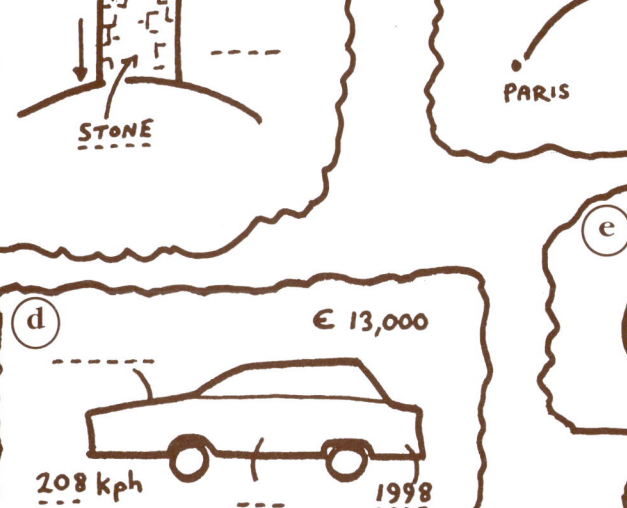

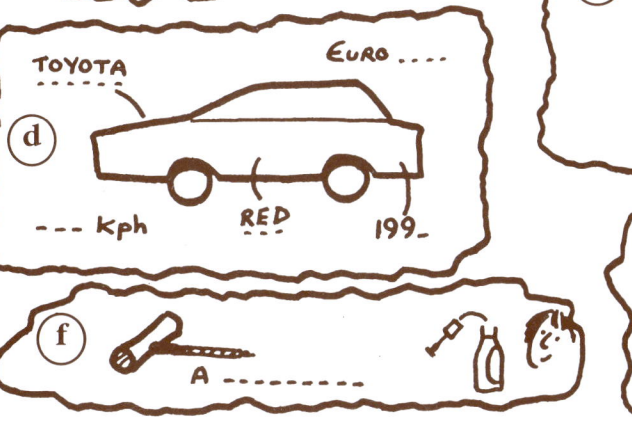
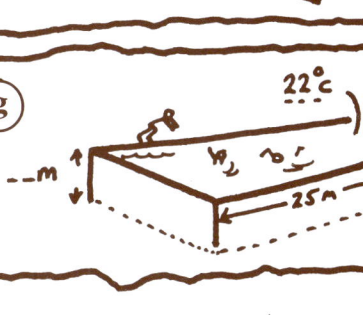

Describing Things – 11A

11B PICTURES OF ENGLISH TENSES LEVEL 2

Describing Things

Use the Picture page 7A to complete the questions.

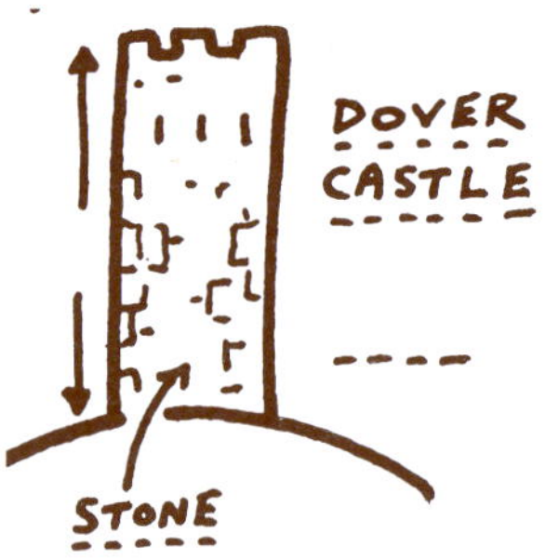

a) What's the name of the castle? — It's Dover Castle.
What's it ....? — It's called Dover Castle.
When ....? — It was built in 1068.
What's it .... of? — It's made of stone.
.... is it? — It's 30m high.

b) How.................. is it from Paris to Moscow? — It's 6440 km
How.................. does it take? — It takes 6 hrs

c) What's .... the jar? — Orange marmalade
How .... is the jar? — It's 450 grammes
How .... is it?
What .... jar is it?
What's ....? — There's a picture of a hand and an orange on the label.

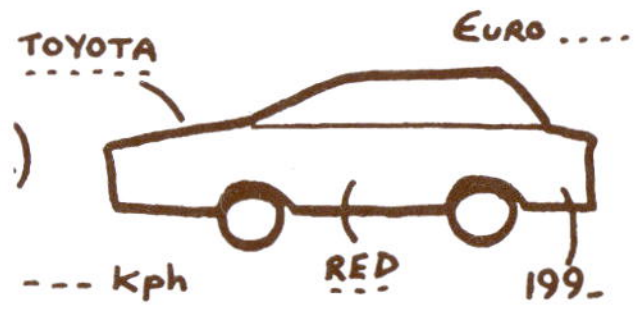

d) What .... / .... of car is it? — It's a Toyota.
What .... is it? — It's red.
How .... is it? — 200 kph.
What's its top .... ?
How .... ? — It costs 13,000 Euros.
What's the .... ? — The price is 13,000 Euros.
When .... it .... ? — It was made in 1998.
How old.... ....?

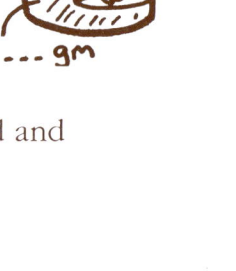

e) How .... is the pipe? — It's 14m long.
What's .... ? — The diameter is 5.5m.
What's .... ? — It's made of steel.
Where .... ? — It was made in Korea.

f) How .... is the pool? — It's 25m long.
How .... is it? — It's 2m deep.
What's .... ? — The temperature is 21deg C.

g) What's this thing .... ? — It's called a 'corkscrew'.
What's it .... ? — It's used for opening bottles.

© English Experience, 25 Julian Road, Folkestone, Kent CT19 5HW

PICTURES OF ENGLISH TENSES LEVEL 2   12A

Revision – 12A

12B PICTURES OF ENGLISH TENSES LEVEL 2

Adriana - The Air Stewardess

Exercise 1

Connect the sentences with the correct picture from Side A.

Picture
- a   Our planes are flown by very experienced pilots.
- b   I've flown to a lot of interesting places.
- c   If I fly for another two years I'll become a senior stewardess.
- d   I fly for Brain-friendly Airways.
- e   In the old days, planes were usually flown by their inventors.
- f   I've been flying for three years.
- g   While we were flying to Sydney, a passenger proposed to me.
- h   Not many planes fly as many passengers as this one.
- i   We're flying over the Andes.
- j   Put your chair in the upright position and fasten your seat belt.
- k   Next month I'm going to fly on the New York City route.

Exercise 2

An onboard interview with Adriana.

Adriana, can you tell me how long you've been flying? 
....
Is there a chance you'll get promotion?
....
Which airline do you work for?
....
This jumbo is massive. How typical is it?
....
What do you say to passengers before landing? 
....
What's the best thing about the job so far?
....
Have you had any funny, or dangerous, experiences?
....
Do you worry about safety in bad weather?
....
Do you know your routes well in advance?
....
Oh, by the way Adriana. Where are we now?
....

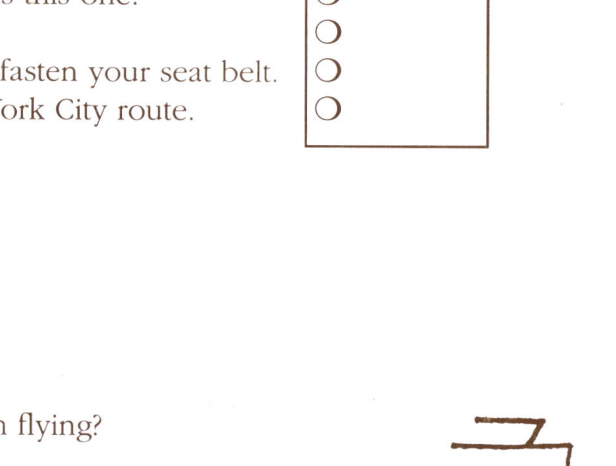

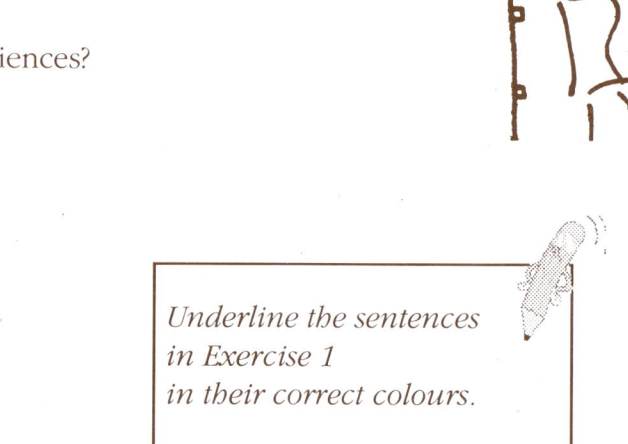

Underline the sentences in Exercise 1 in their correct colours.

© English Experience, 25 Julian Road, Folkestone, Kent CT19 5HW